Growing Up

Caterpillar to Butterfly

by Stephanie Fitzgerald

EXPLORE The LIFE CYCLE!

Content Consultant

Rick Mikula, Butterfly Advisor

D1318704

SCHOLASTIC

Library of Congress Cataloging-in-Publication Data
Names: Fitzgerald, Stephanie, author.
Title: Caterpillar to butterfly / Stephanie Fitzgerald.
Description: New York: Children's Press, an imprint of Scholastic Inc,
   2021. | Series: Growing up | Includes index. | Audience: Ages 6-7. |
   Audience: Grades K-1. | Summary: "Book introduces the reader to the life
   cycle of a butterfly"— Provided by publisher.
Identifiers: LCCN 2020031768 | ISBN 9780531136973 (library binding) | ISBN 9780531137086 (paperback)
Subjects: LCSH: Butterflies—Life cycles—Juvenile literature. |
   Caterpillars—Juvenile literature.
Classification: LCC QL544.2 .F578 2021 | DDC 595.78/9156—dc23
LC record available at https://lccn.loc.gov/2020031768

Produced by Spooky Cheetah Press. Book Design by Kimberly Shake.
Original series design by Maria Bergós, Book&Look.

Printed in Heshan, China 62

SCHOLASTIC, CHILDREN'S PRESS, GROWING UP™, and associated logos are trademarks and/or registered trademarks of Scholastic Inc.

1 2 3 4 5 6 7 8 9 10 R 30 29 28 27 26 25 24 23 22 21

Scholastic Inc., 557 Broadway, New York, NY 10012.

Photos ©: 1 grass and throughout: Freepik; 5 top right: Jannoon028/Getty Images; 6: Adrian Davies/Minden Pictures; 9: Mitsuhiko Imamori/Minden Pictures; 10: Ed Reschke/Getty Images; 12-13 stem: Martin Shields/Science Source; 16 bottom left: K Jayaram/Science Source; 16 bottom right: Steve Kozar/EyeEm/Getty Images; 17 eggs: Jouan Rius/Minden Pictures; 21: SKLA/Getty Images; 22-23 swallowtails: Alberto Ghizzi Panizza/Biosphoto/Minden Pictures; 24: Wild Horizons/Universal Images Group/Getty Images; 25 center right: Jannoon028/Getty Images; 26 bottom left: Brown Bear/Windmill Books/Universal Images Group/Getty Images; 27 top left: Patti Murray/age fotostock; 27 top right: Michael Weber/imageBROKER/age fotostock; 27 center left: Reiner Conrad/Dreamstime; 27 bottom right: Andre Skonieczny/imageBROKER/age fotostock.

All other photos © Shutterstock.

Table of Contents

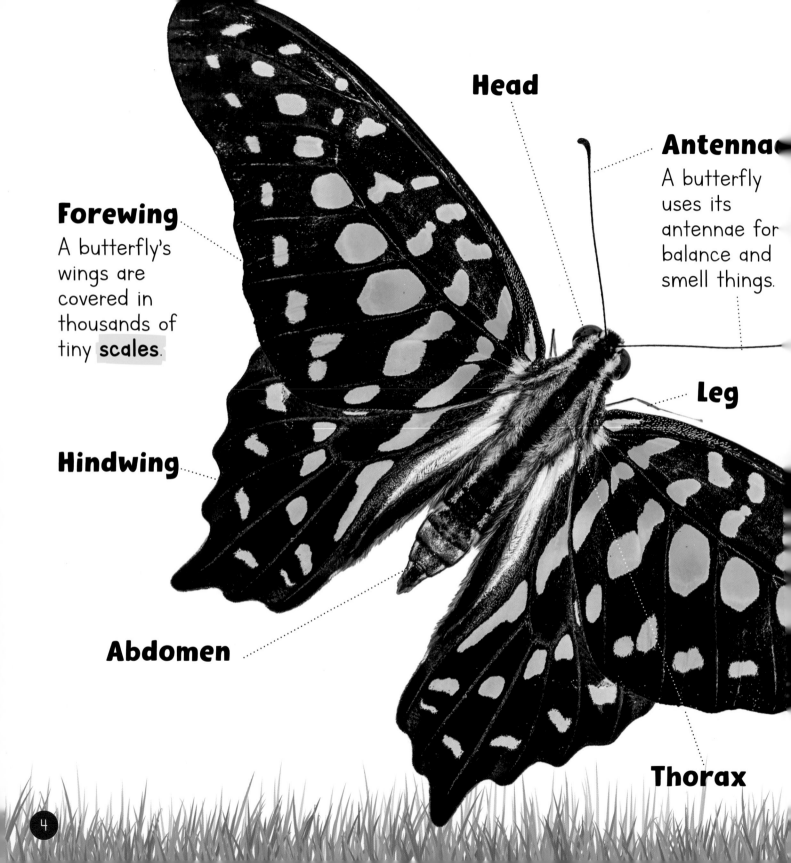

Head

Antennae
A butterfly uses its antennae for balance and smell things.

Forewing
A butterfly's wings are covered in thousands of tiny **scales**.

Leg

Hindwing

Abdomen

Thorax

What Is a Butterfly?

A butterfly is an insect. It has three main body parts: head, thorax, and abdomen. It has six legs, four wings, and two antennae (an-TEN-ee). But a butterfly looks very different at the beginning of its life! Butterflies change as they grow. That process is called **metamorphosis**.

A butterfly has an exoskeleton. Its skeleton is on the outside!

Butterfly eggs can be different shapes and colors. A Malay lacewing lays yellow eggs. ▶

A butterfly makes a sticky fluid. It glues the eggs in place.

It Starts with an Egg

There are many different types of butterfly. But the life cycle of every **species** starts out the same way: with an egg. The female butterfly lays her eggs on the leaves of a plant. The butterfly is very careful about which plant she chooses. The leaves of that plant will be food for her baby.

What's Inside?

Some species lay one egg at a time. Others lay their eggs in bunches. Inside the egg is the butterfly **larva**. It is a tiny caterpillar. The caterpillar is ready to hatch after a few days. It bites through the egg and crawls out onto the leaf.

A caterpillar has 4,000 muscles in its body. People have about 650.

For many caterpillars, their eggshell is their first meal.

▼

Two monarch
caterpillars munch
on the leaves of a
milkweed plant.

A caterpillar may
spend its whole life
on the same plant.

The Hungry Caterpillar

The caterpillar eats lots and lots of leaves. It grows quickly. But the caterpillar's skin does not grow. Soon it starts to feel too tight. The caterpillar must make larger skin. The caterpillar grows new skin under its old skin. Then it sheds the old skin. This is called **molting**.

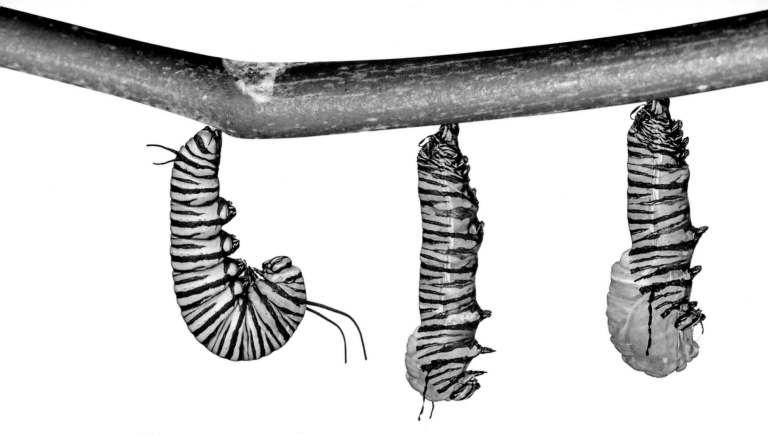

Time to Change

The caterpillar keeps eating and growing. It molts two or three more times. Then, depending on the species, the caterpillar attaches itself to either the stem or the underside of a leaf. It molts one last time.

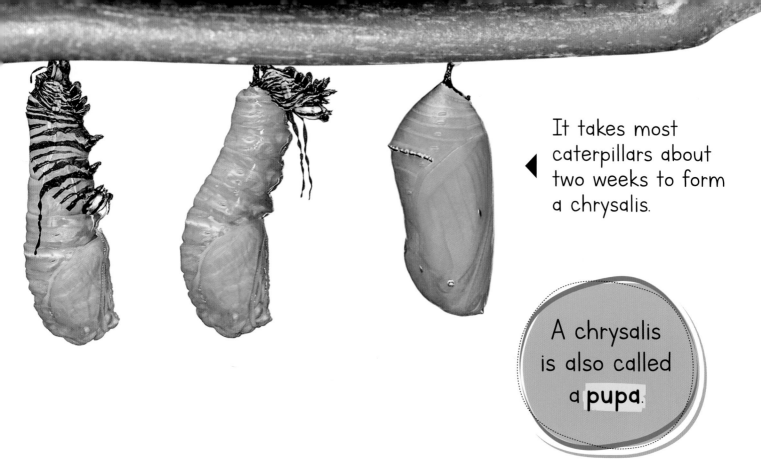

It takes most caterpillars about two weeks to form a chrysalis.

A chrysalis is also called a **pupa**.

This time, something different happens. The caterpillar's new skin forms a hard outer shell called a chrysalis (KRIS-uh-lis). The caterpillar does not eat any more. Now it has important work to do!

When the chrysalis becomes see-through, the butterfly is ready to come out.

Most butterflies spend 10 to 14 days inside the chrysalis.

Breaking Out

Things look quiet outside the chrysalis. But there is a lot happening inside! The butterfly's wings, mouth, antennae, and legs are developing! When the chrysalis finally breaks open, an adult butterfly comes out. The butterfly's wings are crumpled and wet. It cannot fly yet. The butterfly must wait for its wings to dry. It also has to exercise its flight muscles. Then the butterfly is ready to flutter away.

Butterfly Predators

The wings of an adult butterfly are usually very colorful. Its scales create beautiful patterns. These patterns can help the butterfly blend into its surroundings. That helps the butterfly stay safe from **predators**. Threats to the butterfly change depending on the stage of its life cycle.

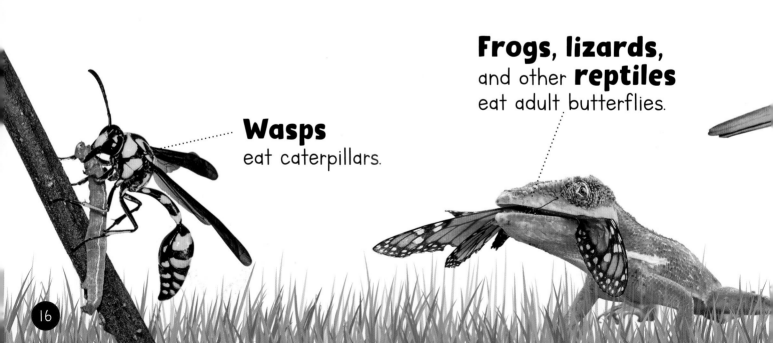

Wasps
eat caterpillars.

Frogs, lizards,
and other **reptiles**
eat adult butterflies.

Some caterpillars look like bird poop. That helps them hide from predators.

Birds
eat caterpillars and adult butterflies.

Ants
eat butterfly eggs.

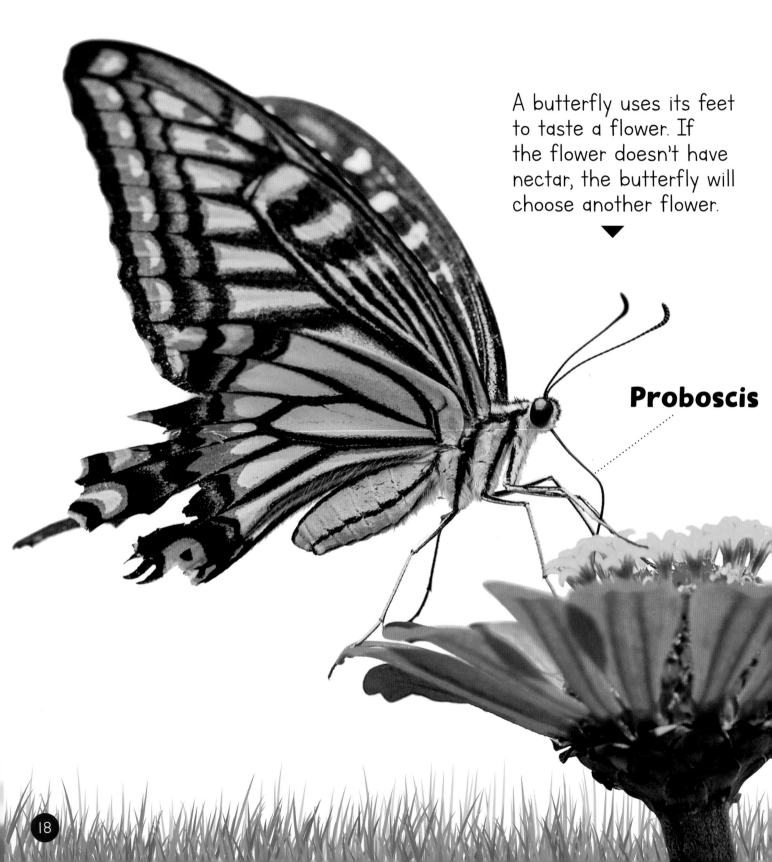

A butterfly uses its feet to taste a flower. If the flower doesn't have nectar, the butterfly will choose another flower.

▼

Proboscis

Sipping Sweet Nectar

The adult butterfly has a sipping mouthpart called a **proboscis.** Butterflies feed on flower nectar. That's where they get the energy to fly! The butterfly uses its proboscis to sip nectar and to drink water.

When a butterfly isn't feeding, its proboscis curls up.

Making New Flowers

Adult butterflies are **pollinators**. Tiny grains of pollen stick to the butterfly as it rests on a flower to eat. As the butterfly flutters from flower to flower, it spreads the pollen around. That helps new flowers grow.

Male flowers make pollen. It is carried to female flowers to make new flowers.

This monarch
butterfly has yellow
pollen on its legs.
▼

Finding a Partner

Soon it is time for the butterfly to find a **mate** of the same species. It searches for a butterfly with the right colors and patterns on its wings. When the butterflies choose each other, they may fly together. It looks like a dance.

Two scarce
swallowtails meet
on a flower.

A Doris longwing lays about 200 eggs at a time.

▼

The Life Cycle Begins Again

Now the female must lay her eggs. She has to find the right plant. She can tell by the shape and color of its leaves. The butterfly may also stomp on the leaves. That releases their scent. Now the butterfly is sure she has found the perfect plant. She lays her eggs. The life cycle begins again.

Butterfly Facts

Butterflies cannot tolerate cold weather. Some species **migrate** to warmer areas in the fall. Monarchs are the migrating champs of the butterfly world. They travel 3,000 miles to escape the cold.

Some butterflies, like the buckeye, have eyespots on their wings. The eyespots are meant to confuse or scare predators.

The female Queen Alexandra's birdwing is the largest butterfly in the world. Its wingspan is about as long as a football!

The western pygmy blue is the smallest butterfly species. Its wingspan is less than an inch across!

Butterflies are known for being colorful. But the glasswing butterfly's wings are see-through!

There are about 17,500 known species of butterflies in the world. The United States is home to 750 species. The cabbage white is the most common.

Most butterflies fly about 5 to 12 miles per hour. But skippers are the fastest. They can reach speeds of 37 miles per hour.

Growing Up from Caterpillar to Butterfly

A butterfly goes through a complete metamorphosis as it grows. It changes from a tiny caterpillar to a beautiful butterfly.

Most butterflies live a few weeks. Brimstone butterflies can live 10 to 14 months.

Eggs
Butterfly eggs are tiny—about as small as the head of a pin.

Larva
A caterpillar hatches from each egg. Caterpillars look nothing like adult butterflies!

Adult
Once it reaches this stage, the butterfly is ready to mate and start the life cycle again.

Pupa or chrysalis
This stage can last from a few days to a few months. Some species of butterfly spend the winter inside the chrysalis.

Glossary

larva (LAHR-vuh) an insect at the stage of development between an egg and a pupa, when it looks like a worm

mate (MAYT) the male or female partner of a pair of animals

metamorphosis (met-uh-MOR-fuh-sis) a series of changes some animals go through as they develop into adults

migrate (MYE-grate) to move to another area or climate at a particular time of year

molting (MOHLT-ing) losing old fur, feathers, shell, or skin to be replaced by new growth

pollinators (PAH-luh-nay-torz) animals that carry or transfer pollen among flowers so that the flowers can produce seeds

predators (PRED-uh-turz) animals that live by hunting other animals for food

proboscis (pruh-BAH-sis) a long straw-like tube that a butterfly uses to eat and drink

pupa (PYOO-puh) an insect in the stage of development between a larva and an adult

scales (SKALZ) the thin, flat, overlapping pieces of hard skin that cover the wings of a butterfly

species (SPEE-sheez) one of the groups into which animals and plants are divided

Index

About the Author

Stephanie Fitzgerald is the author of more than 30 books for kids. She's very curious about the world around her, which is why she likes writing nonfiction—there's always something new to discover! Fitzgerald lives in Connecticut with her family and lots of butterfly-friendly plants.